DAVID DONG
BIOLOGY 1A RM L 103 LEC
L 112 LAB
LENN, BOB

The Fetal Pig

A PHOTOGRAPHIC STUDY

W. L. Evans
The University of Arkansas

Addison E. Lee
The University of Texas

Photographs by **George Tatum**
The University of Texas

REVISED EDITION

Holt, Rinehart and Winston
NEW YORK—CHICAGO—SAN FRANCISCO
TORONTO—LONDON

April, 1966

Printed in the United States of America
Library of Congress Catalog Number: 58-7998
22592-0118

Preface

This book is designed as a visual aid for students in the biological sciences — biology, zoology, nursing, comparative anatomy — who use the fetal pig as a laboratory animal for dissection and study.

The student should follow the instructions for the dissection of his specimen and use the labeled photographs to help him recognize the various structures encountered in the dissection. Not only can the student locate more easily the structures which are to be studied, but he will often avoid the unnecessary cuts and mutilation of his specimen which could make the identification of structures to be studied later difficult or impossible.

The photographs have been grouped according to the organ systems and each of the important structures has been numbered; in some instances the number is on the structure, and in others it connects to an arrow which points to the structure. Labels corresponding to the numbers are given on the page opposite each photograph. The same structures are sometimes labeled on different photographs in order to identify them from different views or to serve as points of reference for the student.

This book will also be helpful to the student in reviewing for examinations, because it will give him a record of the structures he has studied. However, it should be used as an aid, not a substitute, for a careful study of the specimen itself. In order to facilitate its use, the book has an alphabetical index which lists all structures illustrated and their page numbers.

The authors are indebted to Dr. Frances Clayton for helpful criticism and suggestions in the preparation of the first edition of the book; to Jerry J. Raisch and David Lacker, R. T., for providing the X ray of the fetal pig on page 9, and to Neta W. Lee for typing the text and preparing the index.

THE AUTHORS

Contents

The Fetus

Fetal pigs are frequently used as laboratory specimens for detailed anatomical study of higher animals. These animals are particularly advantageous in this respect because they belong to the highest class of vertebrates, the mammals; they are relatively inexpensive, easy to obtain in quantity, and are easily stored. Unlike cats or rabbits, pigs are seldom pets and the average student will undertake dissection of the fetal pig more readily than some other animals. The fetal pig is an unborn animal and thus the bones and other tissues are not completely hardened, which makes dissection easier.

Fetal pigs usually are obtained from slaughter houses. The illustrations below show a uterus which had been removed from the body of a sow that had been butchered for meat. The illustration on the left shows the fetus removed from the uterus, but still within the chorioallantoic membrane (fetal placenta). The illustration on the right shows the fetus removed from the chorioallantoic membrane. This fetus is somewhat younger than those generally used by students for dissection, as will be illustrated on the following pages.

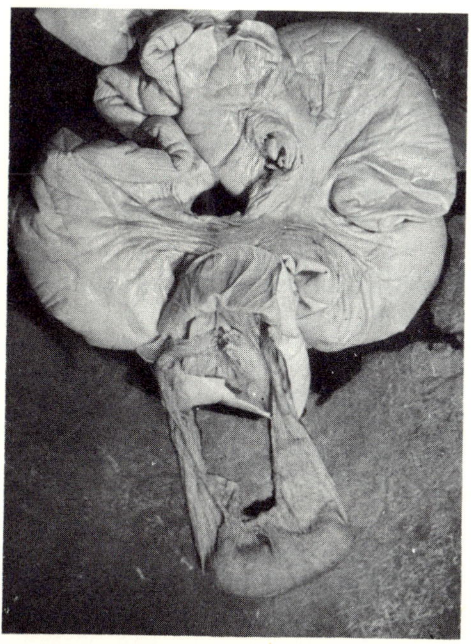

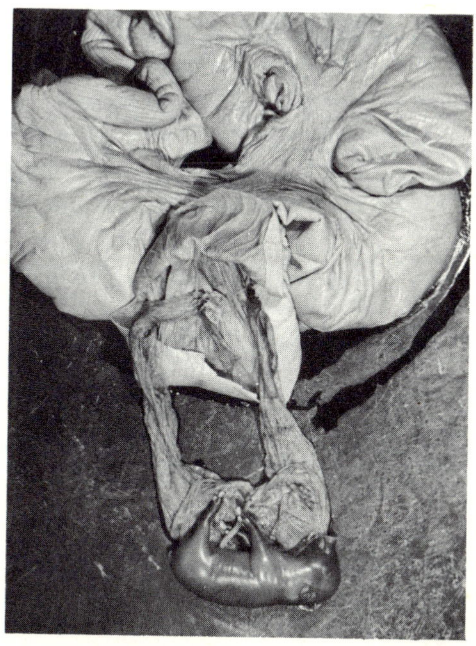

I. External Structures

Begin your study with a thorough examination of the external characteristics and structures. Note each structure or region on the picture, then identify it on your specimen.

THE FETAL PIG

A Anterior end
P Posterior end
D Dorsal side
V Ventral side

BODY REGIONS

M *Median* or midline through the body as opposed to the sides (*lateral*). A dorsoventral cut parallel to the median line reveals a *sagittal* section. A *transverse* section would be one at right angles to the long axis of the body.

1 Head
2 Cervical region (neck)
3 Thoracic region
4 Lumbosacral region
5 Caudal region

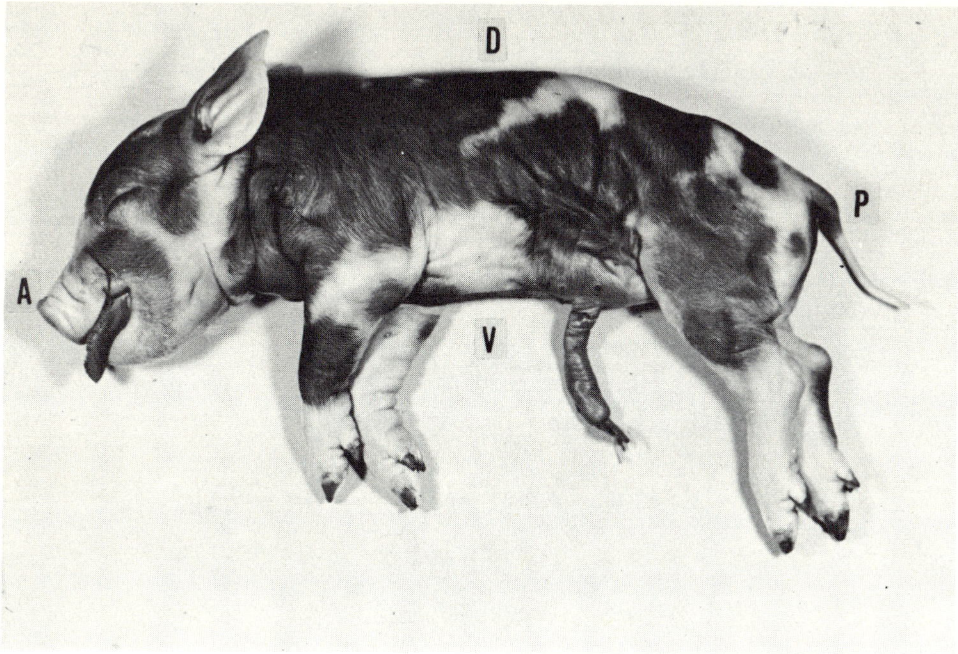
D
A
P
V

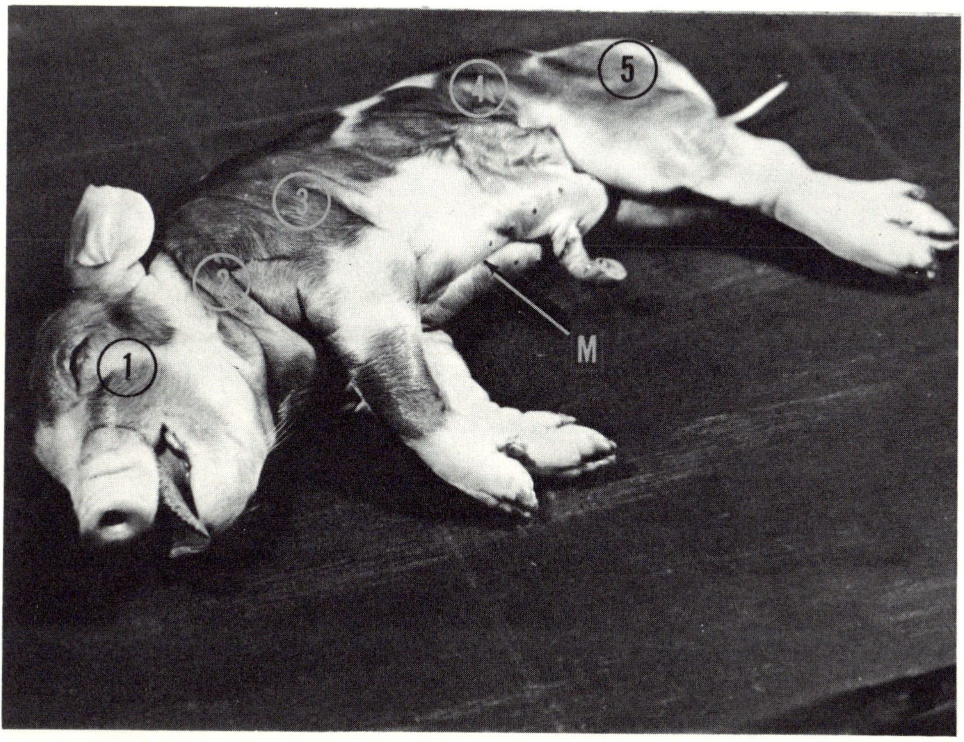
1
2
3
4
5
M

EXTERNAL ANATOMY

6 External nare (nostril)
7 Forelimb
8 Hoof
9 Umbilical cord
10 Hind limb

HEAD, EXTERNAL ANATOMY

11 Tongue
12 Nictitating membrane
13 Eye
14 External auditory aperture
15 Pinna

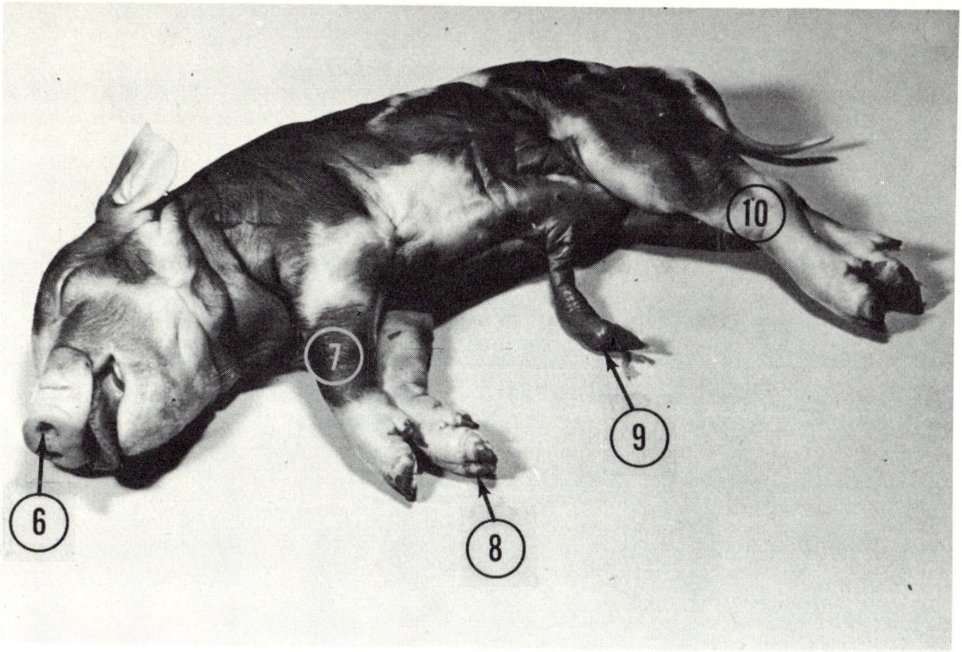
6
7
8
9
10

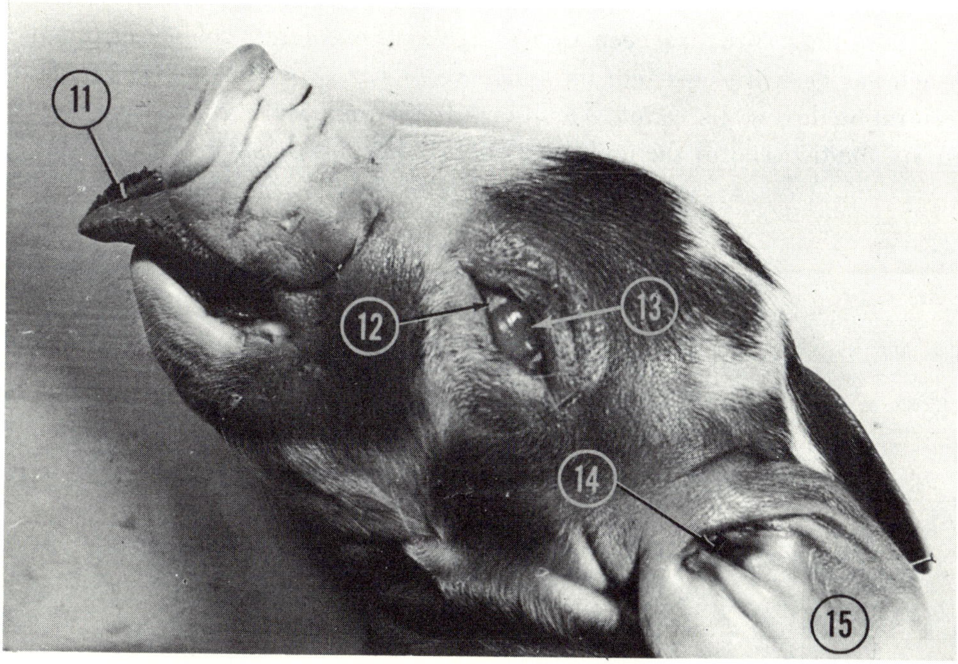
11
12
13
14
15

VENTRAL POSTERIOR EXTERNAL ANATOMY, MALE

16 Opening of penis (male urogenital opening)
17 Teat
18 Umbilical cord
19 Scrotum

The umbilical cord has been tied to retain the colored rubber compound which has been injected into the blood vessels in order to prevent their deterioration during dissection. To observe these umbilical vessels listed below, cut off the tied end of the umbilical cord about half an inch from the tie.

VENTRAL POSTERIOR EXTERNAL ANATOMY, FEMALE

20 Teat
21 Umbilical vein
22 Umbilical arteries

The female urogenital opening is located just ventral to the anus and is not shown in this illustration.

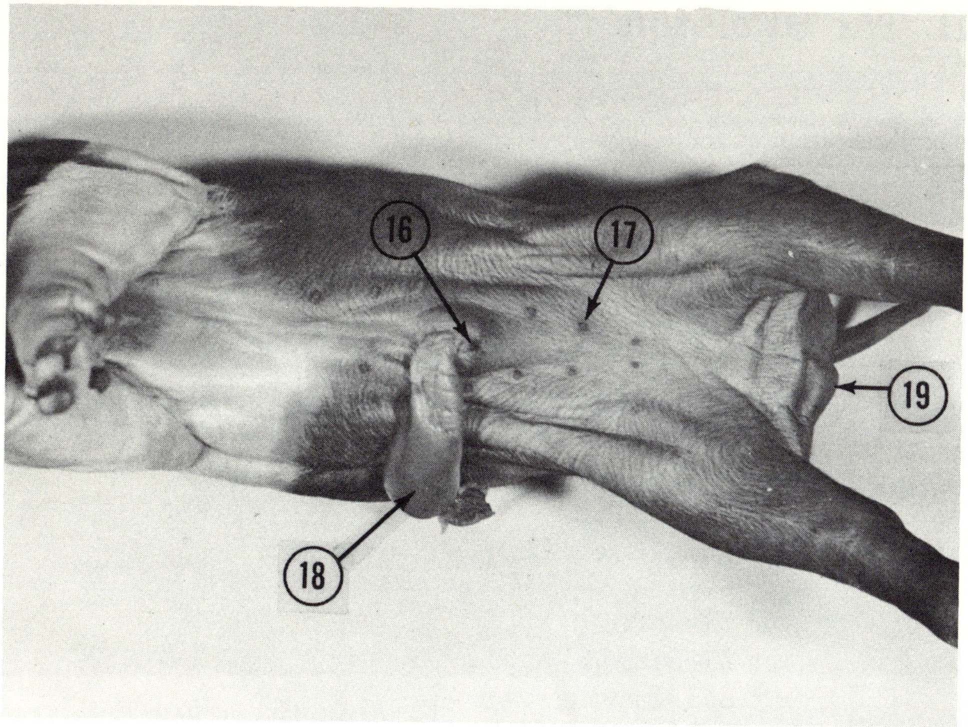
16
17
18
19

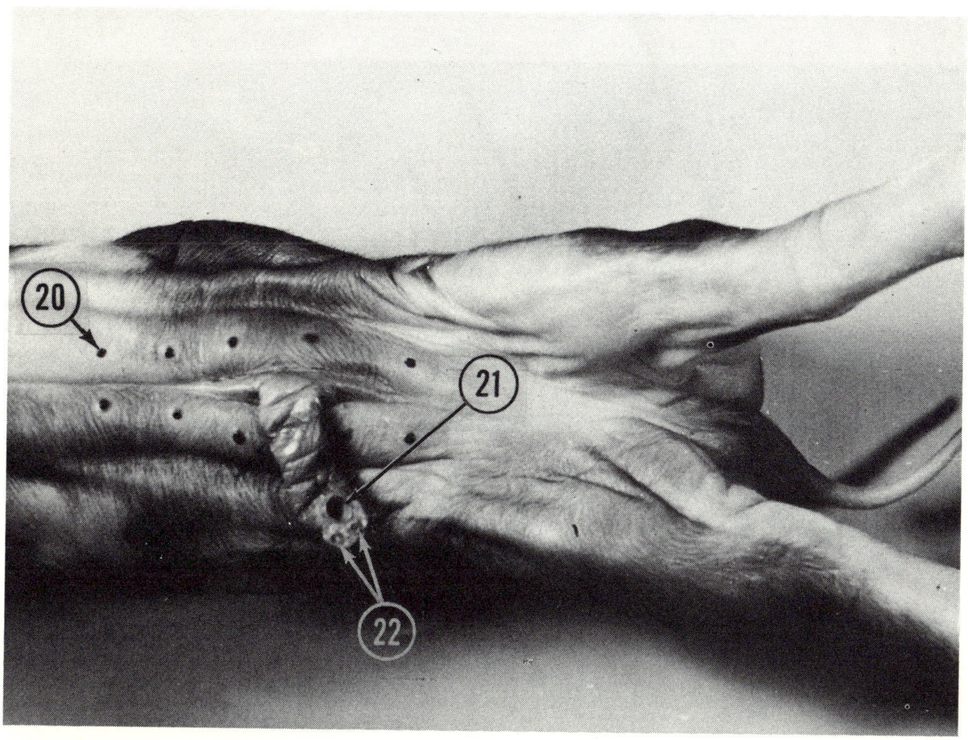
20
21
22

II. The Skeleton

The illustration on the following page was made from an X ray of the fetal pig. The visceral organs and the fore- and hind limbs from one side were removed before the X ray was taken.

Although detailed dissection and study of the complete skeleton of the fetal pig are not very practical for the beginning sudent, it is nonetheless necessary for him to become somewhat acquainted with some of the more important bones and their general arrangement prior to his dissection of the various internal organs which are to be studied in detail. A study of the illustration on the opposite page and the labels below should accomplish this purpose.

23 Skull
24 Cervical vertebrae
25 Thoracic vertebrae
26 Lumbar vertebrae
27 Sacral vertebrae
28 Caudal vertebrae
29 Lower jaw
30 Humerus
31 Radius
32 Ulna
33 Bones of the forefoot
34 Sternum
35 Rib
36 Pelvis
37 Femur
38 Tibia (note fibula underneath)
39 Bones of the hind foot

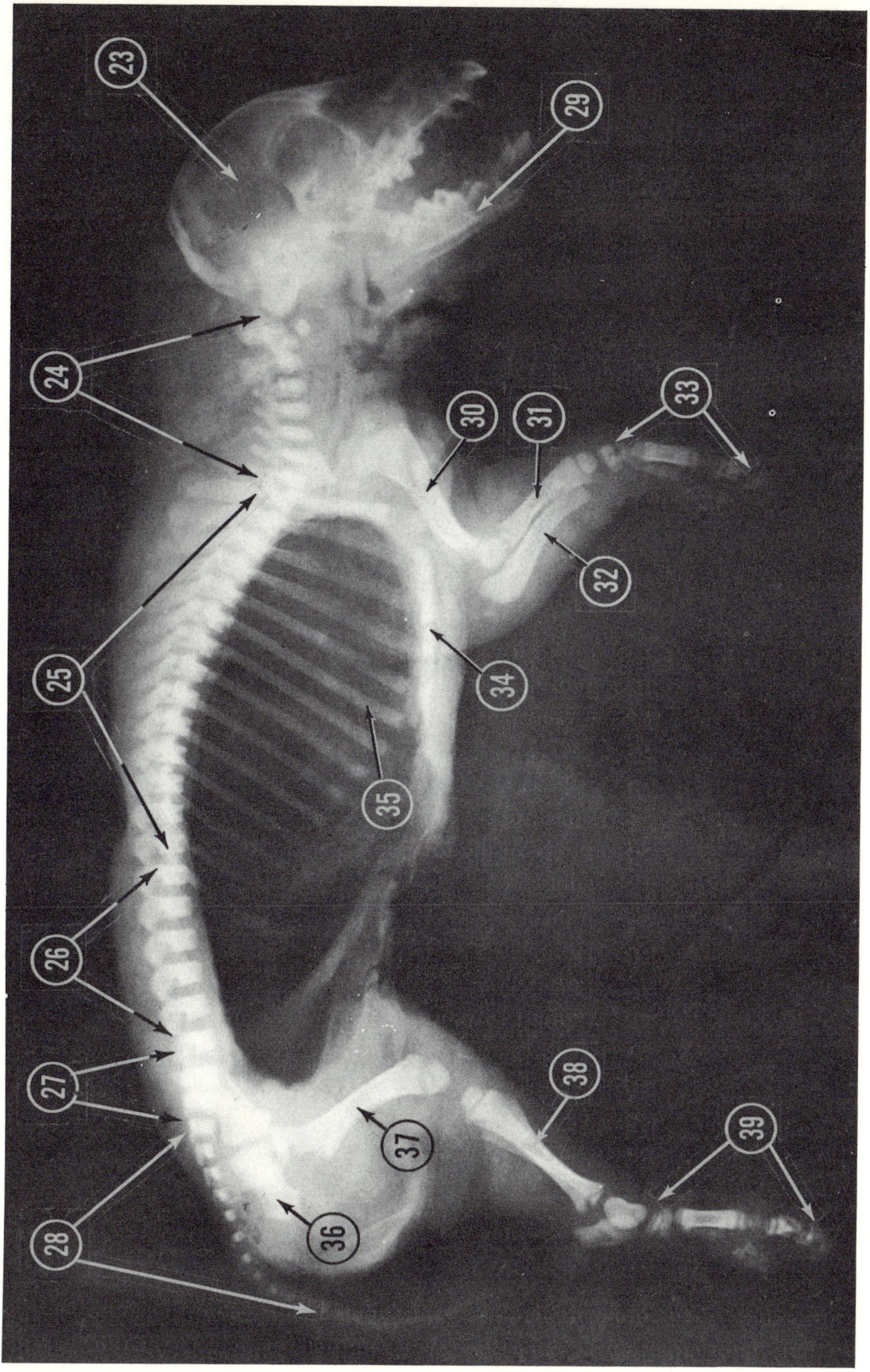
23
24
25
26
27
28
29
30
31
32
33
34
35
36
37
38
39

III. Mouth and Pharynx

Using scissors, cut posteriorly about an inch at the angle made by the joint of the upper and lower jaws. When bone is felt, use bone shears for cutting through the bone. After the cut has been made on both sides, hold the snout portion and press downwardly on the tongue and lower jaw in order to fully expose the mouth cavity. It may be necessary to break and tear muscles at the corners of the mouth.

MOUTH HELD OPEN *jaw muscles and bones have been partially cut and broken*

40 External nare
41 Glottis
42 Epiglottis

MOUTH HELD OPEN

43 Hard palate (note ridges)
44 Soft palate
45 Tongue

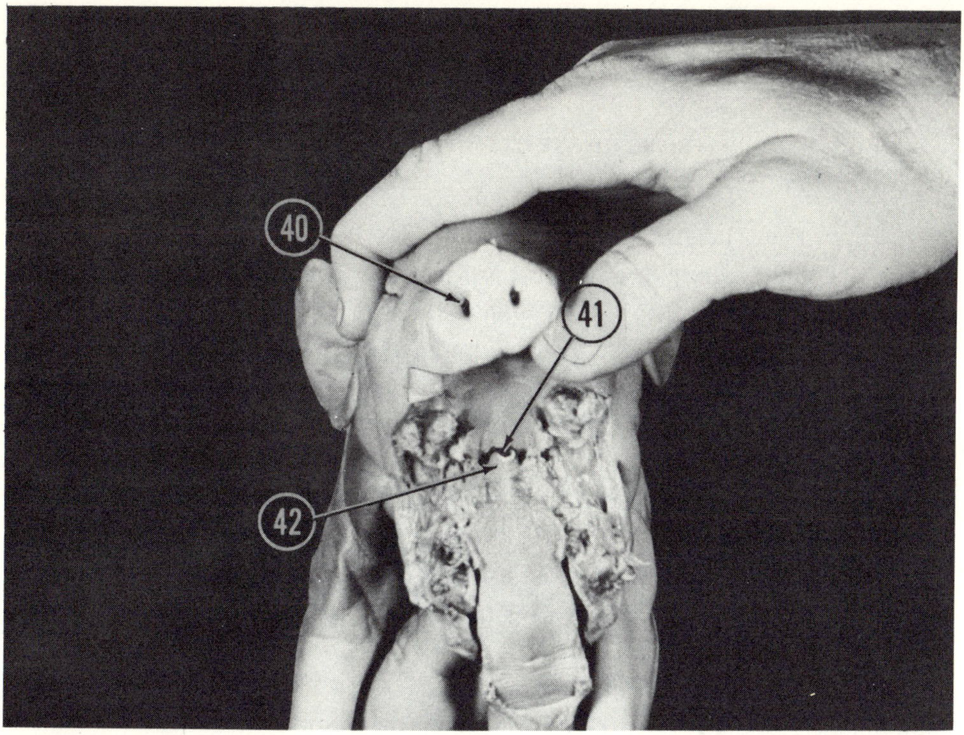
40
41
42

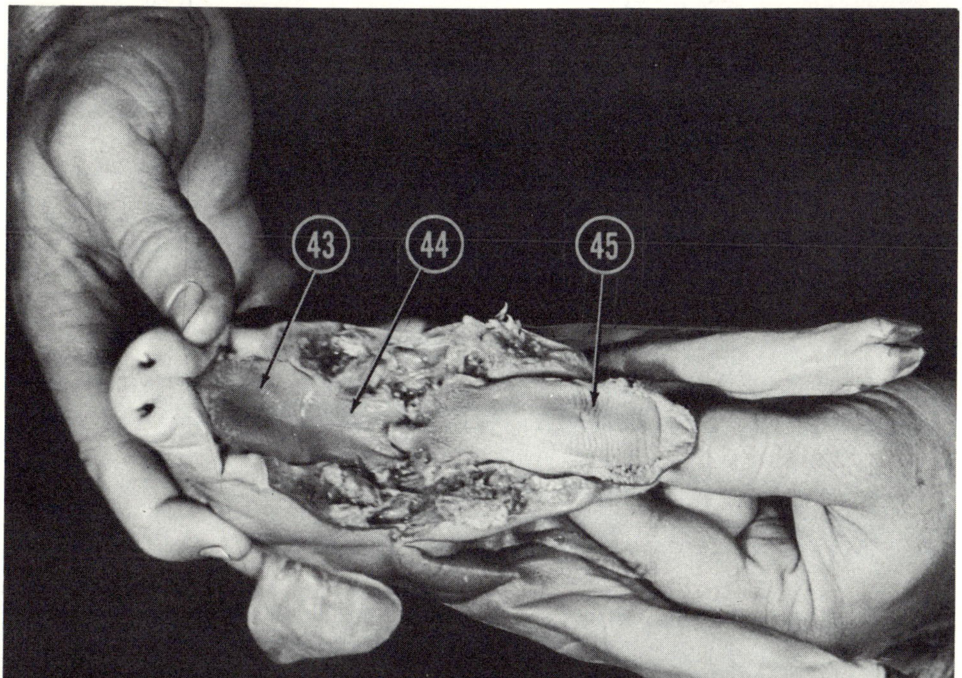
43
44
45

IV. Neck Organs

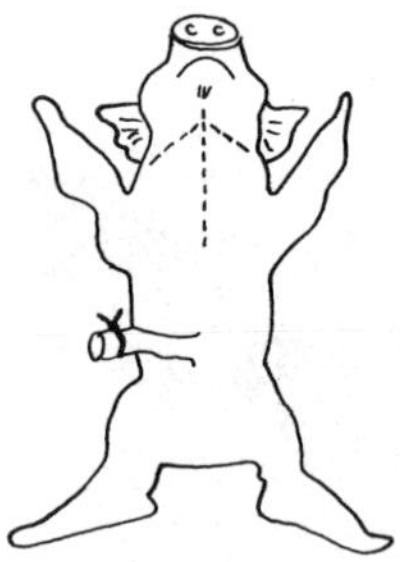

THROAT DISSECTION

Place the animal on its back in the dissecting pan and locate the neck region. Carefully cut through the skin along the ventral midline, and extend the cut from the clump of hairs under the chin (see the photograph) to an inch or so behind the forelegs. With a blunt instrument, such as a probe or the handle of the scalpel, separate the skin from the underlying muscles. Now cut the skin transversely for about two inches on each side of the neck in order to lay the skin back. With a scalpel, cut through the tissue in the neck region and separate the muscles and connective tissue along the midline by pulling these laterally. This exposes the thymus gland.

46 Skin
47 Muscles and connective tissue
48 Thymus gland (also extends into the thoracic cavity)

THROAT DISSECTION *with thymus gland pulled laterally*

With forceps, pick away the connective tissue lying along the midline between the two parts of the thymus gland, which is in the neck region, and push this tissue laterally. Identify the organs lying underneath.

49 Larynx
50 Trachea
51 Thymus gland
52 Thyroid gland

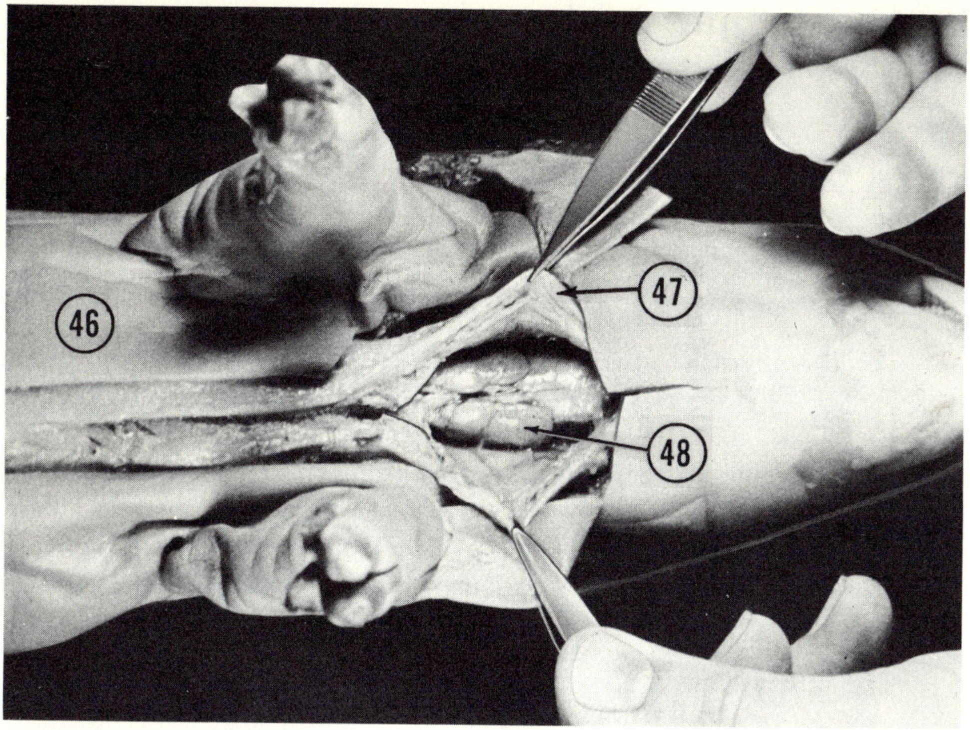
46
47
48

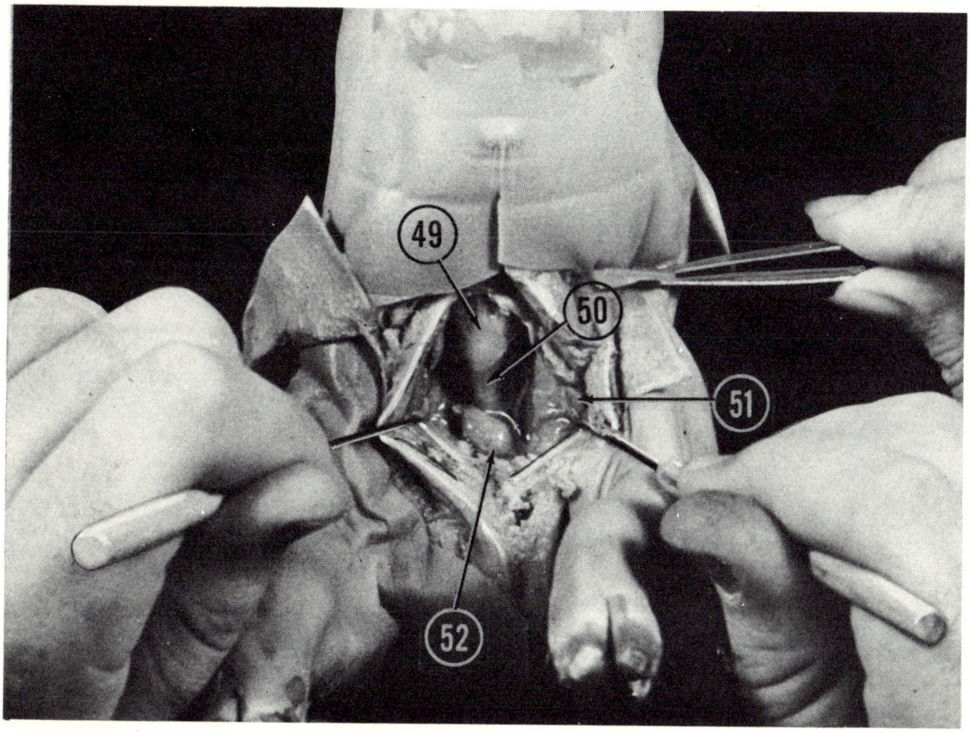
49
50
51
52

V. Abdominal Cavity

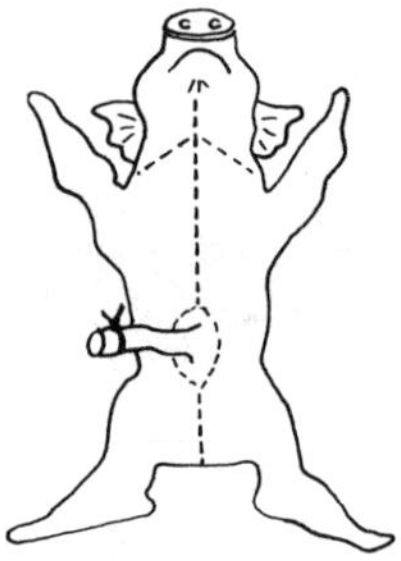

Continue the initial cut through the skin posteriorly to about an inch anterior to the base of the umbilical cord. Extend the cut around the umbilical cord on both the left and the right side. Continue a single incision through the skin only posterior to the umbilical cord. This single incision should be in the midline and should begin posterior to the urogenital opening in the male. Separate the skin from the underlying muscles and repeat the cuts through the abdominal wall. (Exercise care posterior to the umbilical cord in the male pig to avoid cutting the penis.) Make two lateral cuts on each side. One should be just posterior to the ribs and the other should be just anterior to the leg (see the photograph). Identify the structures listed below.

ABDOMINAL DISSECTION *with umbilical cord in natural position*

53 Umbilical cord
54 Umbilical vein
55 Penis
56 Umbilical artery
57 Liver
58 Spleen

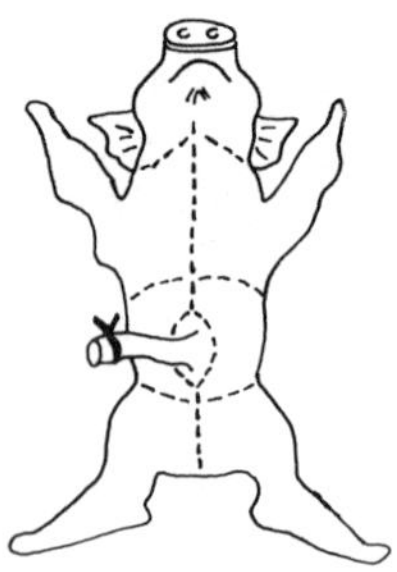

Tie the umbilical vein (54) in two places for later identification. Cut between the two ties and lay the umbilical cord between the hind legs. Proceed with identification of the various structures.

ABDOMINAL DISSECTION *with umbilical vein severed and umbilical cord pulled posteriorly*

59 Spleen
60 Liver
61 Large intestine
62 Small intestine
63 Umbilical arteries

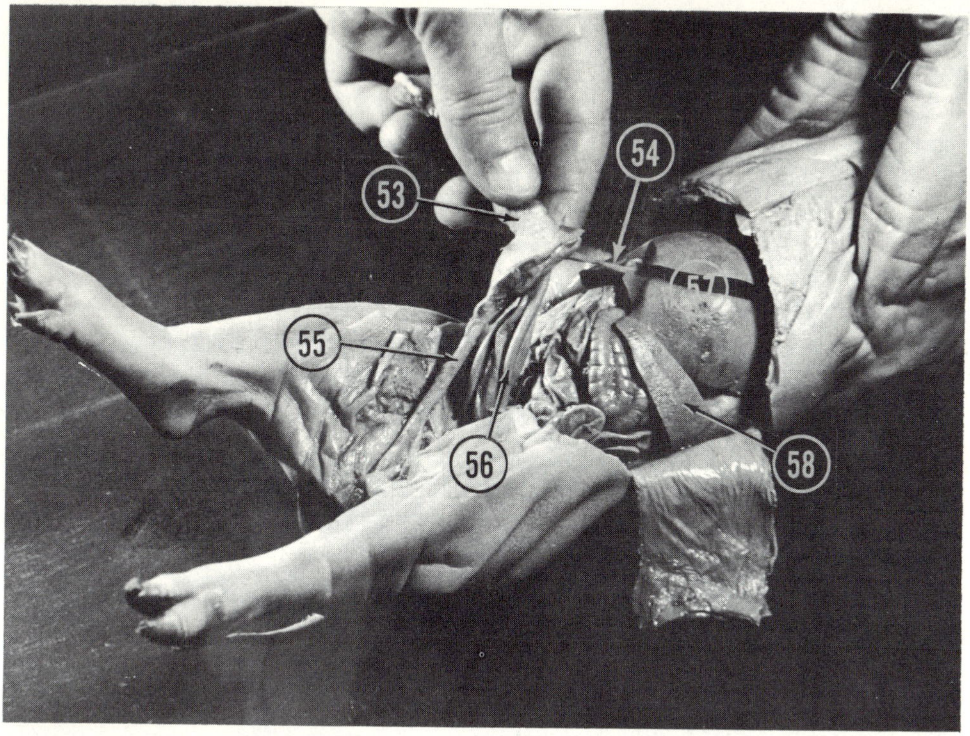
53
54
57
55
56
58

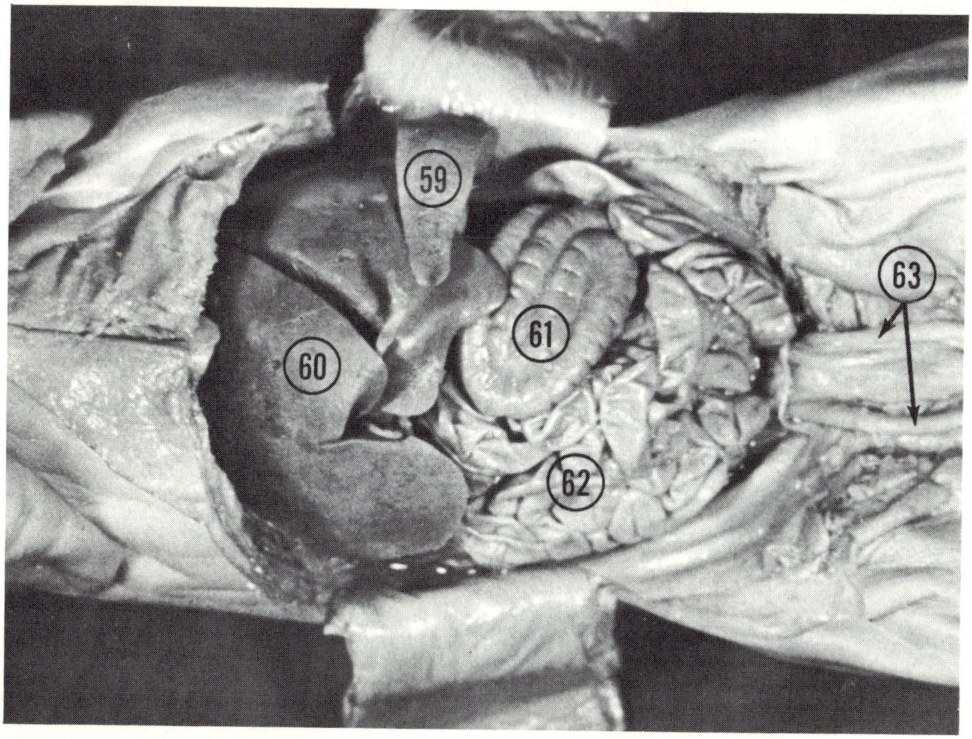
59
63
61
60
62

In studying the organs of the abdominal cavity, it is necessary to displace some of the organs to gain access to others. For example, the gall bladder lies underneath the liver, and the latter organ must be pulled forward to expose the former.

ABDOMINAL DISSECTION *with liver raised*

64 Spleen
65 Mesentery (greater omentum)
66 Gall bladder
67 Bile duct

ABDOMINAL DISSECTION *with intestine pulled laterally*

68 Caecum
69 Large intestine
70 Mesentery
71 Small intestine

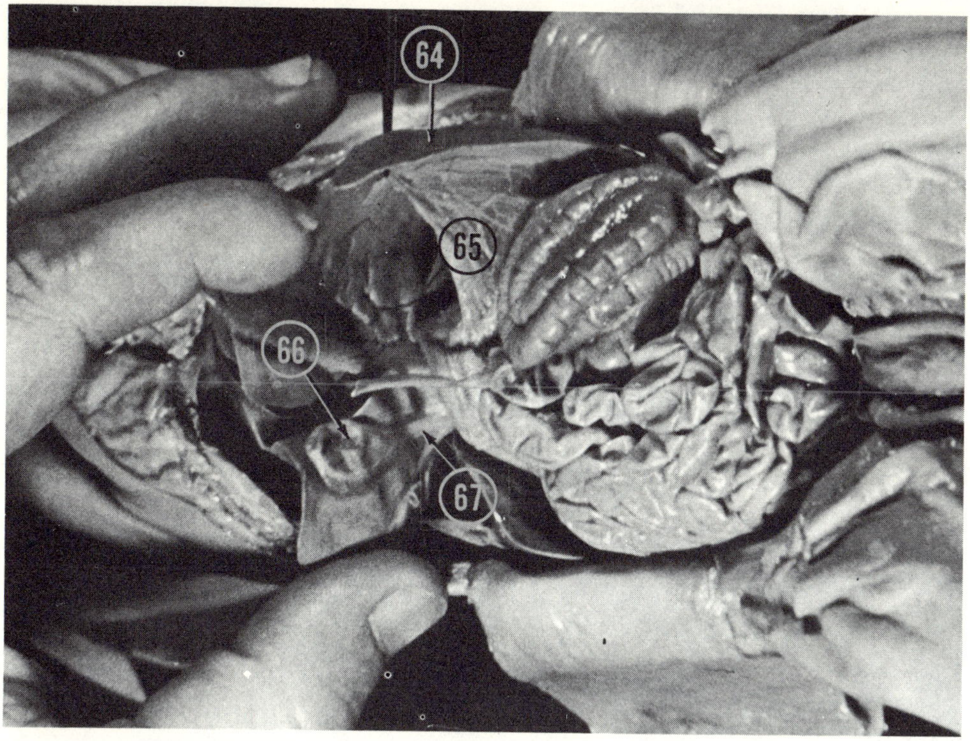
64
65
66
67

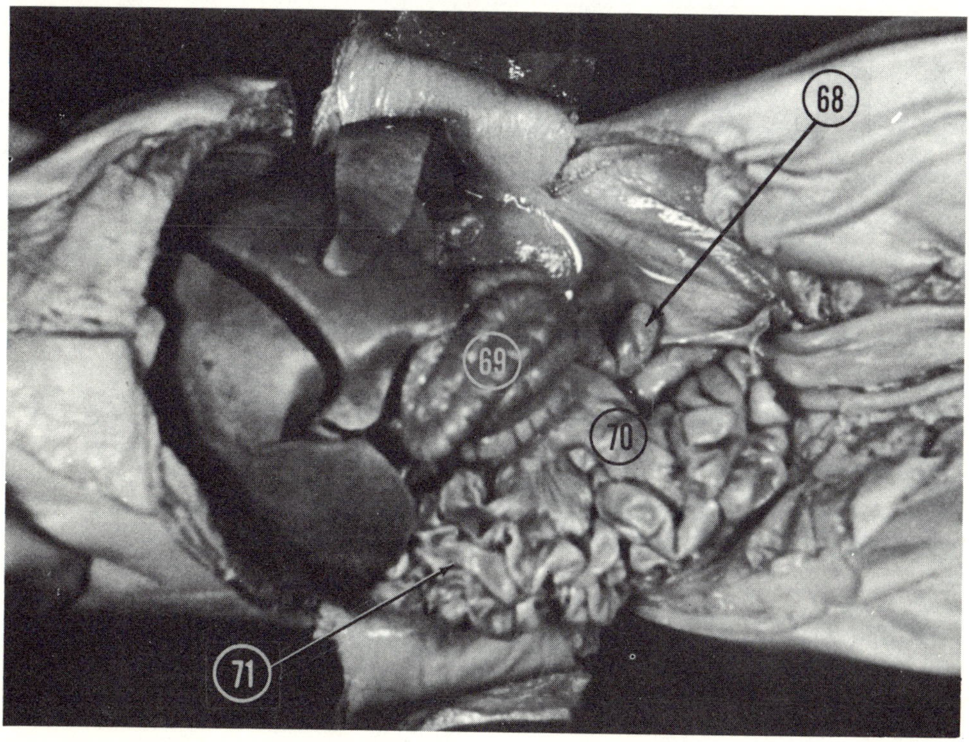
68
69
70
71

ABDOMINAL DISSECTION *with abdominal organs pulled arteriorly and laterally*

72 Stomach
73 Adrenal gland
74 Kidney
75 Pancreas
76 Large intestine
77 Rectum

ABDOMINAL DISSECTION *with intestines pulled laterally*

78 Liver
79 Large intestine
80 Caecum
81 Small intestine
82 Rectum

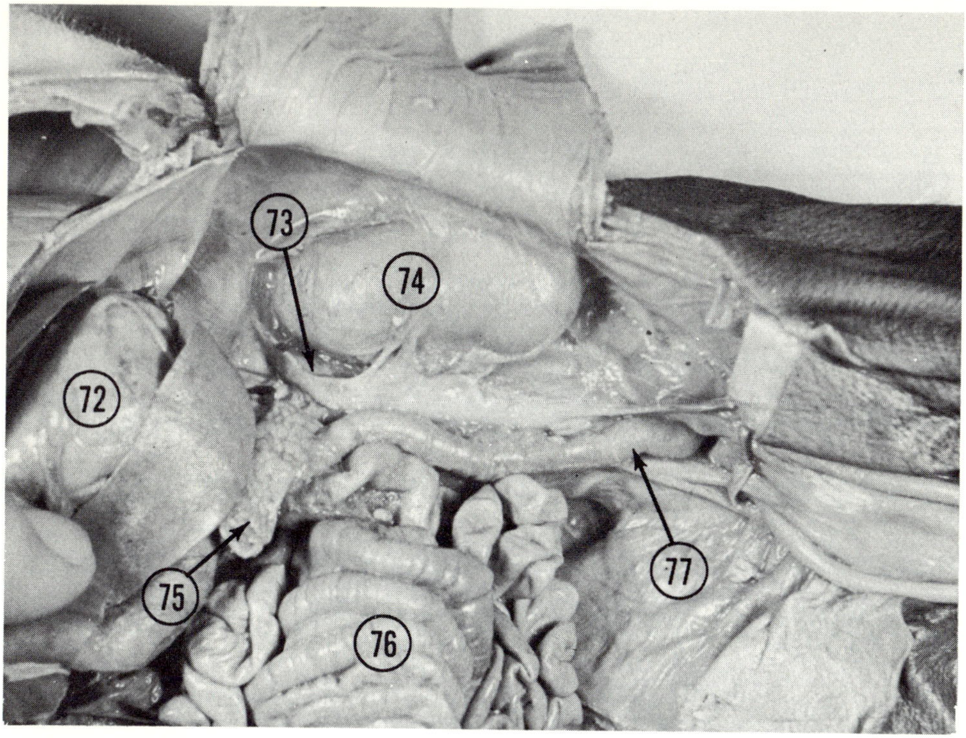
72
73
74
75
76
77

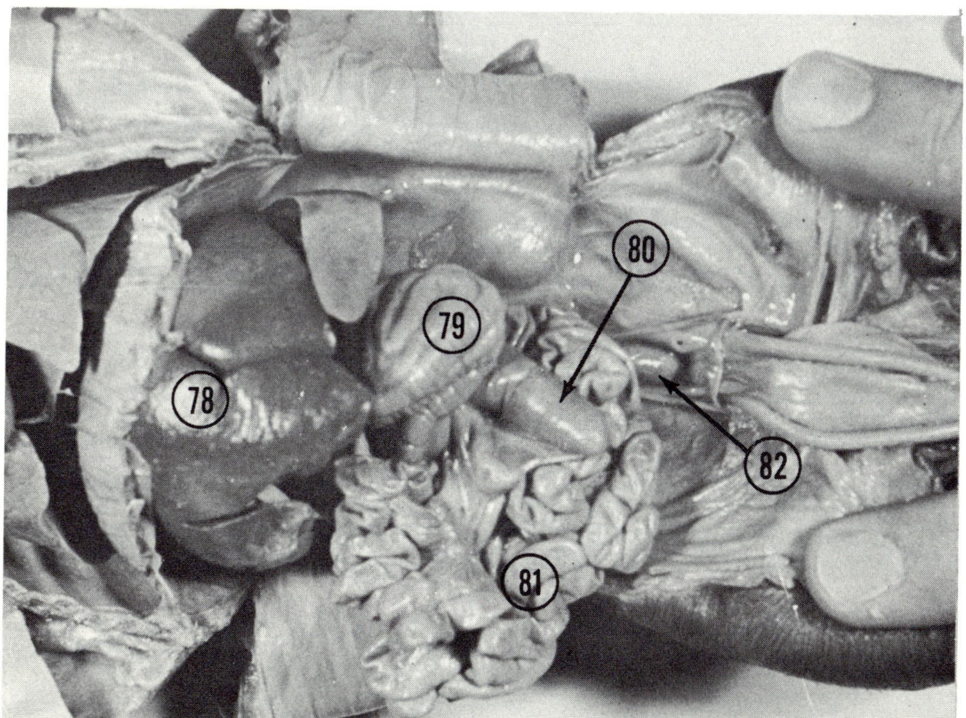
78
79
80
81
82

ABDOMINAL DISSECTION *with intestines pulled anteriorly*

83 Rectum
84 Umbilical artery
85 Inguinal canal
86 Urinary bladder

Note the shiny surface anterior to the fingers. This is the *peritoneum.*

ABDOMINAL DISSECTION *with liver removed and spleen pulled laterally*

87 Stomach
88 Cardiac region
89 Pyloric region
90 Pancreas
91 Pyloric sphincter
92 Duodenum (upper part of small intestine)

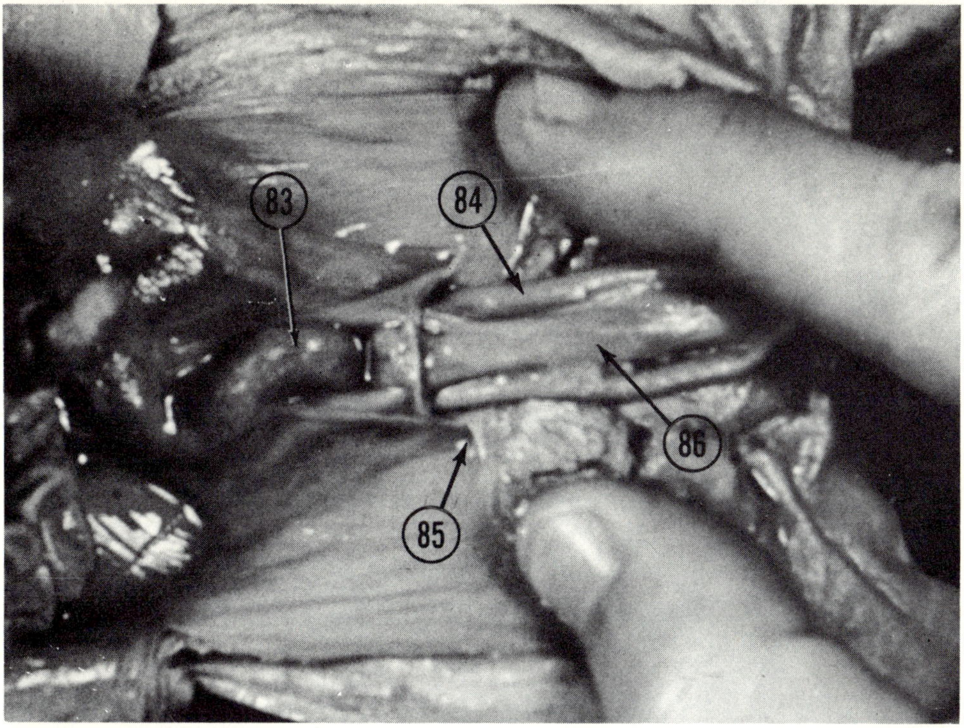
83
84
86
85

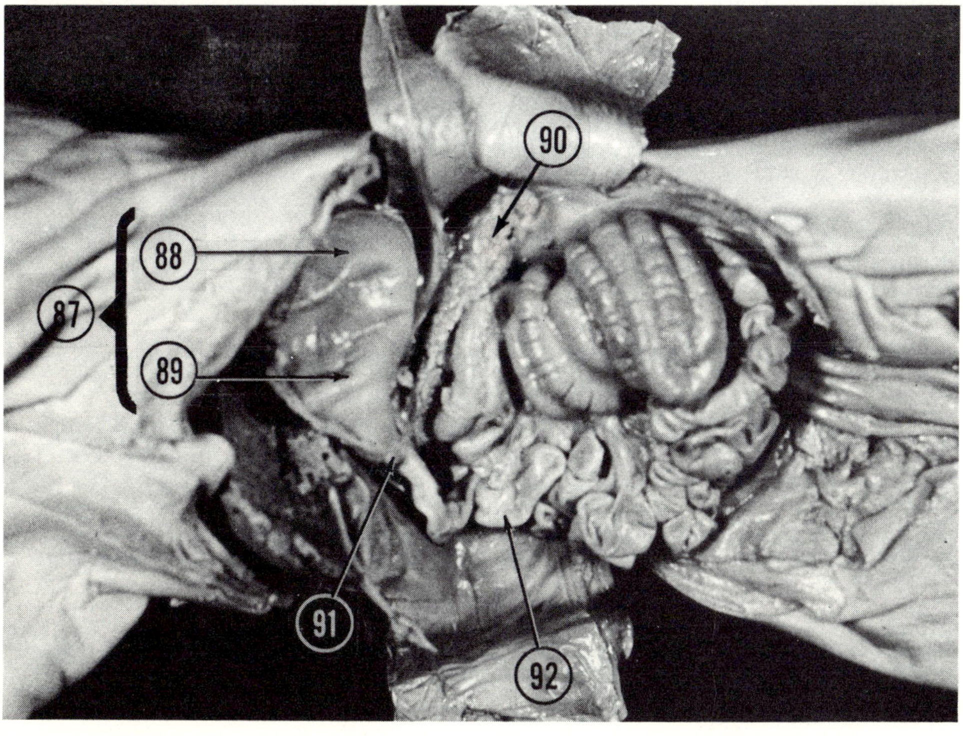
90
88
87
89
91
92

VI. Urogenital Organs

Remove the abdominal organs which have already been studied. Locate the scrotum and extend the incision along the midventral line to cut through the scrotum. Care must now be exercised in cutting deeper. Preserve the tube-like structure (penis) as you make these cuts. Continue to cut and gently separate the muscles and other tissue in this region until each scrotal sac is evident. Forceps should be used in picking away the connective tissue. Identify the structures listed below.

UROGENITAL DISSECTION, MALE *with intestines removed and peritoneum partially removed*

93 Kidney
94 Ureter
95 Adrenal gland
96 Urinary bladder
97 Rectum (cut)
98 Genital artery

Open one scrotal sac for examination.

UROGENITAL DISSECTION, MALE

99 Seminal vesicle
100 Scrotal sac
101 Prostate
102 Spermatic cord
103 Testis
104 Gubernaculum

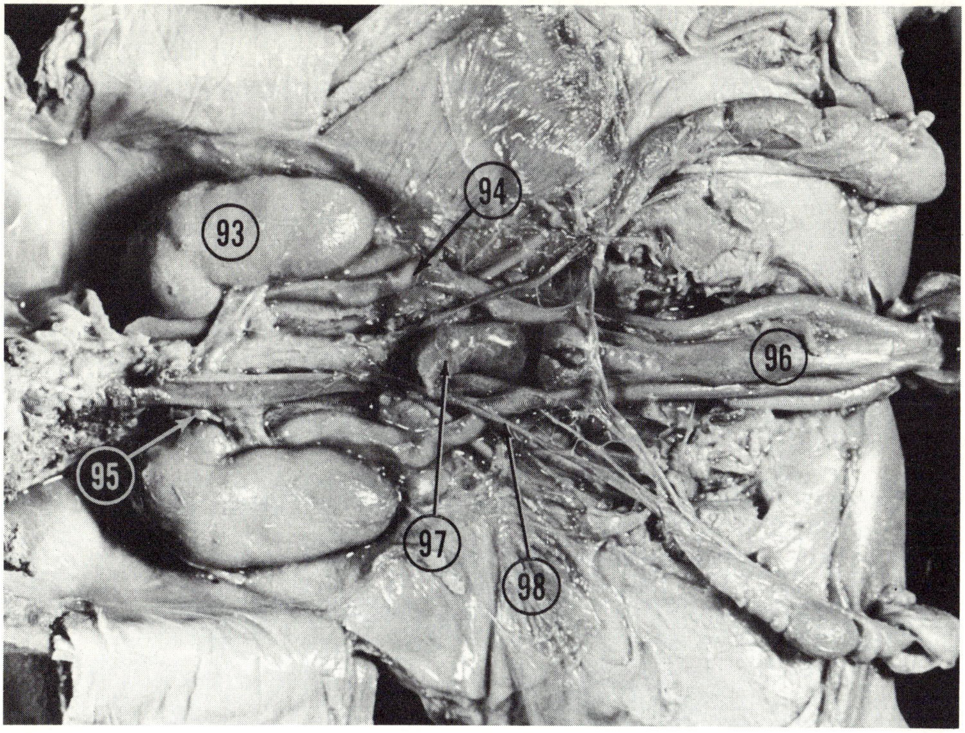
93
94
96
95
97
98

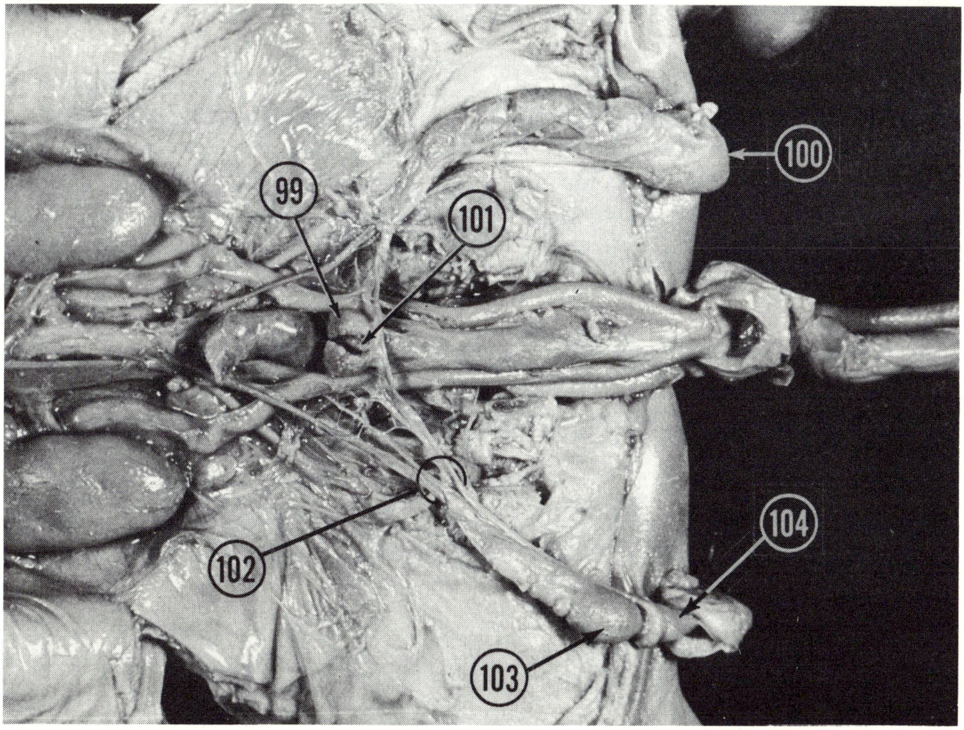
100
99
101
104
102
103

While holding the bladder and umbilical arteries laterally, carefully cut through the pelvic bone slightly to one side of the midline. In order to do this, force the legs laterally so they are flattened against the dissecting board. This should expose the male urogenital tract from the base of the bladder where the urethra arises to the external urogenital opening. Pick the connective tissue away with forceps and identify the structures listed below.

UROGENITAL DISSECTION, MALE *with pelvic girdle cut*

105 Penis
106 Urinary bladder
107 Urethra
108 Bulbo-urethral gland (Cowper's gland)
109 Vas deferens
110 Epididymis

Identification of the female structures can begin when the abdominal organs have been removed.

UROGENITAL DISSECTION, FEMALE

111 Kidney
112 Fallopian tube
113 Ovary
114 Ostium
115 Horn of the uterus
116 Urinary bladder

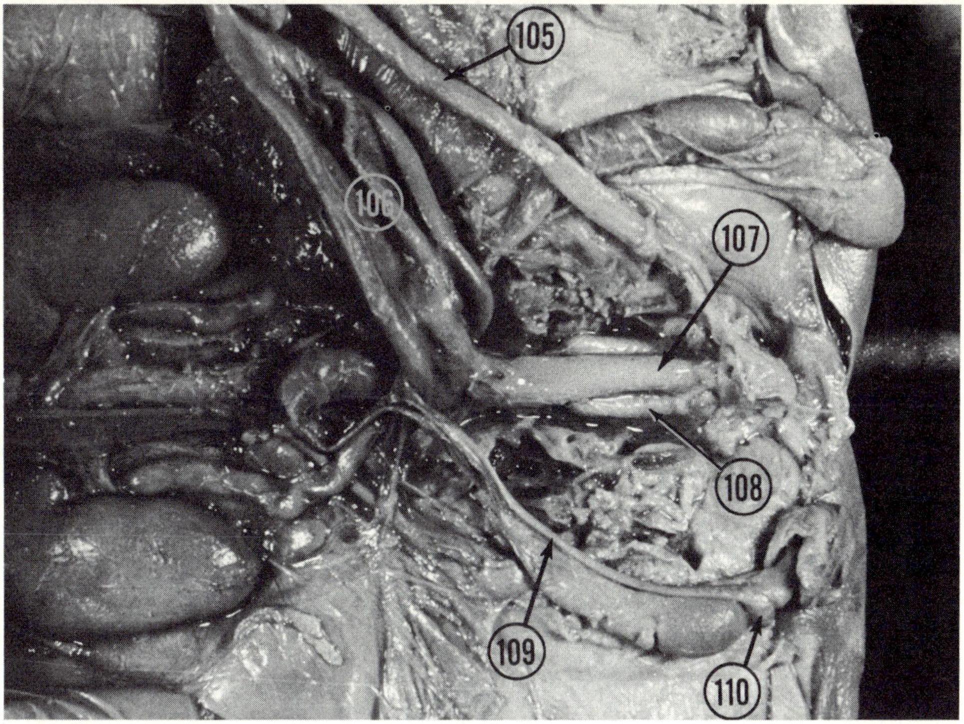
105
106
107
108
109
110

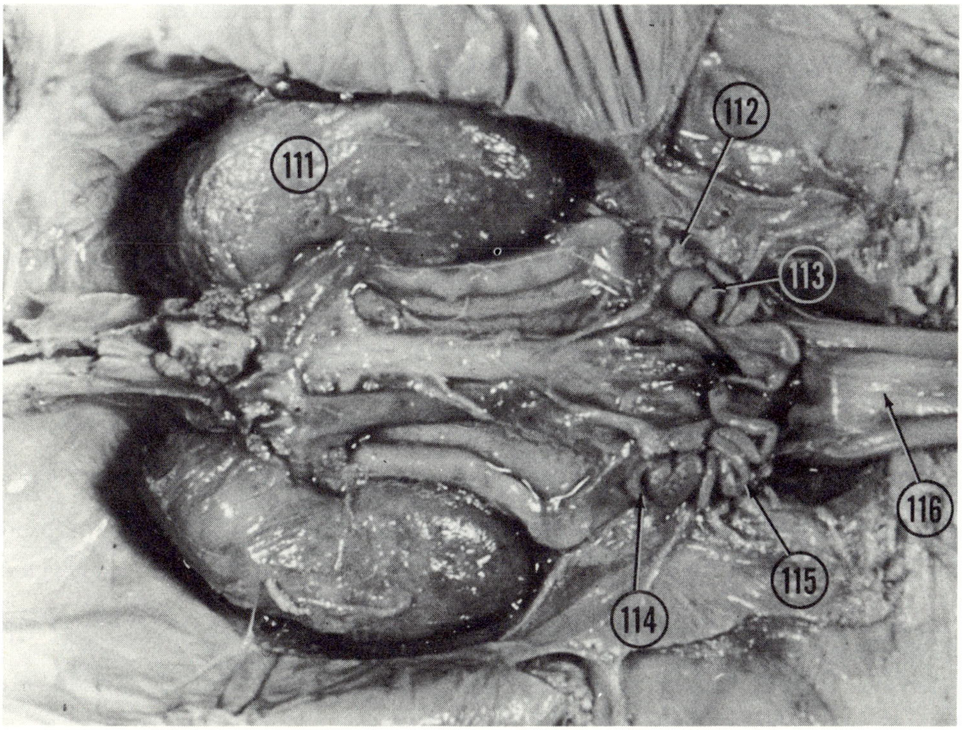
111
112
113
114
115
116

Pull the urinary bladder and umbilical vessels laterally and cut through the pelvic bone, which is found about half an inch below the base of the bladder. The cut should be made slightly to one side of the midline, and the connective tissue and muscles around the female urogenital organs should be picked away carefully with forceps. Identify the structures listed.

UROGENITAL DISSECTION, FEMALE *with urinary bladder pulled laterally and the pelvic girdle cut*

117 Ureter
118 Uterus (partly obscured by the umbilical artery)
119 Vagina
120 Urethra
121 Umbilical artery
122 Urinary bladder

UROGENITAL DISSECTION, FEMALE *with left ureter and umbilical artery cut and urinary bladder pulled laterally*

123 Ovary
124 Horn of the uterus
125 Uterus
126 Vagina
127 Urogenital sinus

Note margin of the rectum lying dorsal to the vagina and urogenital sinus.

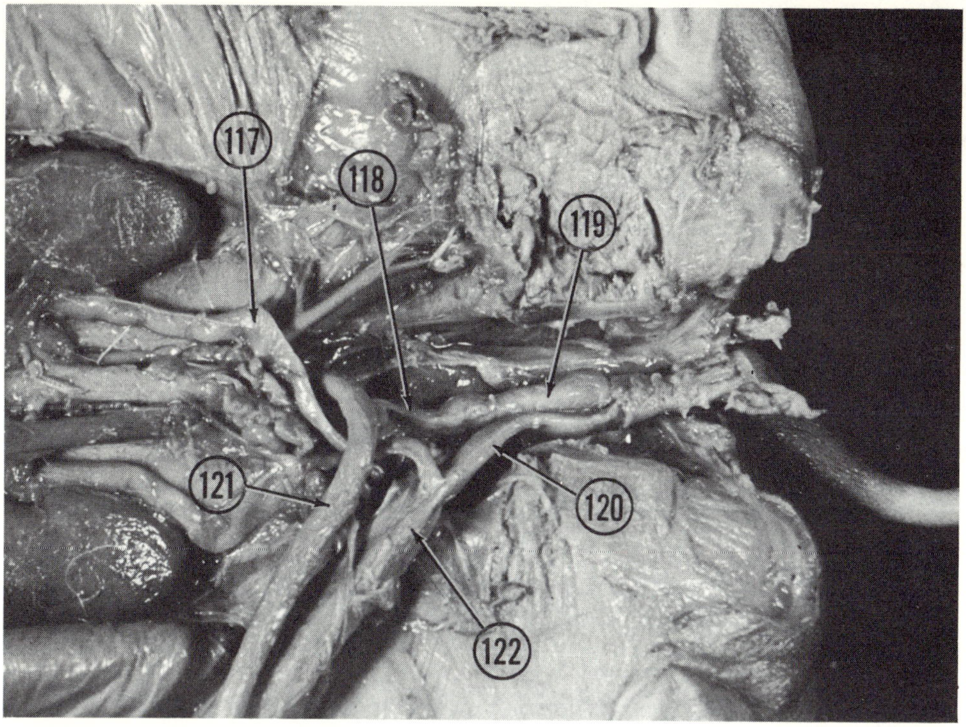
117
118
119
121
120
122

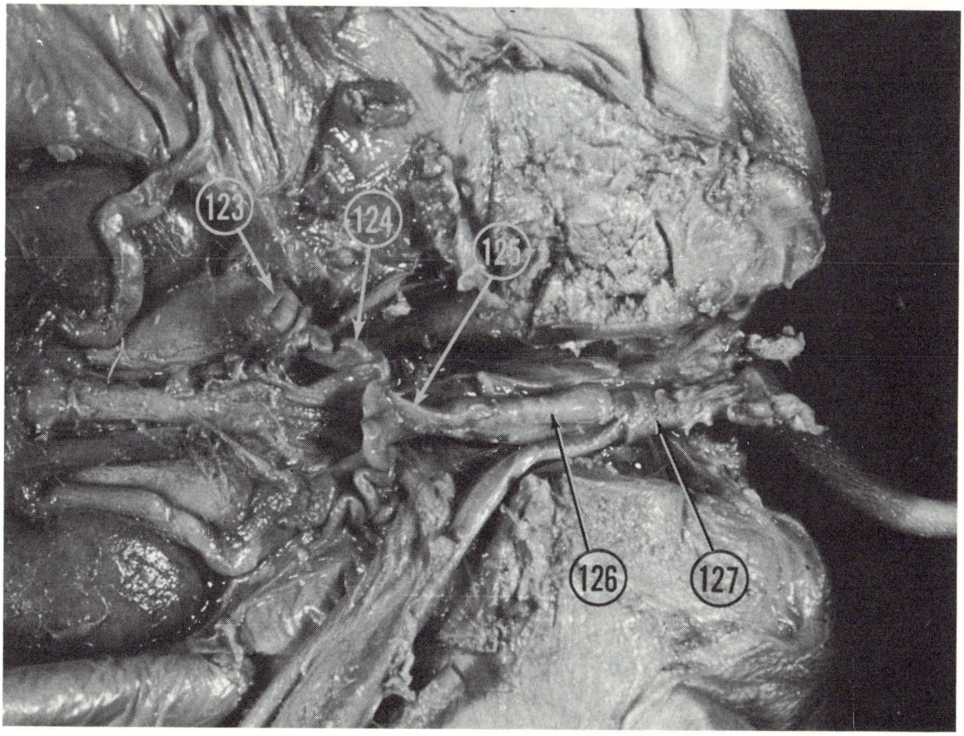
123
124
125
126
127

VII. Thoracic Cavity and Heart

Insert scissors at the junction of the sternum and the diaphragm, and make an incision along the midventral line in the thoracic region. Cut forward to a point midway between the forelegs. Care should be exercised to avoid cutting too deeply. Now cut along the junction of the ribs and the diaphragm on each side so the entire thoracic cavity is fully exposed. Identify the organs listed.

THORACIC DISSECTION *with sternum and part of the ribs removed*

128 Portion of the thymus gland
129 Pericardium
130 Heart
131 Lung (the membrane which lines the lung cavity is called the *pleura*)
132 Ribs
133 Diaphragm

Note the neck organs previously identified anterior to the thoracic cavity. The thymus gland extends into the thoracic cavity. Note that the visceral organs posterior to the thoracic cavity have been removed.

THORACIC DISSECTION *with pericardium and thoracic portion of thymus gland removed*

134 Left auricle
135 Right auricle
136 Coronary blood vessels
137 Left ventricle
138 Apex of the heart
139 Right ventricle

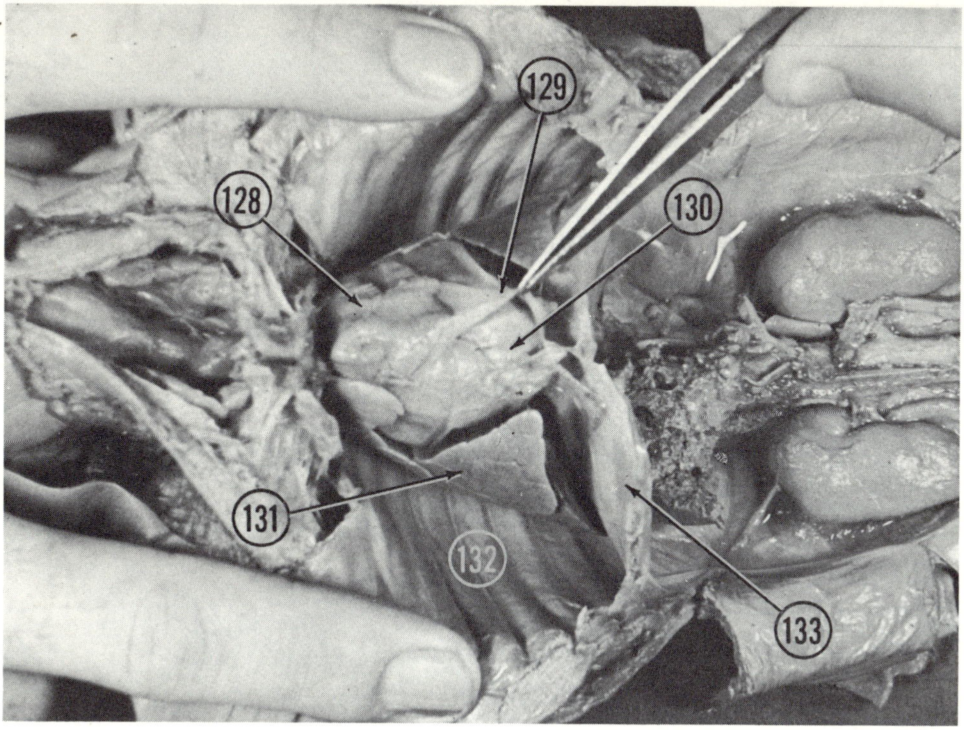
128
129
130
131
132
133

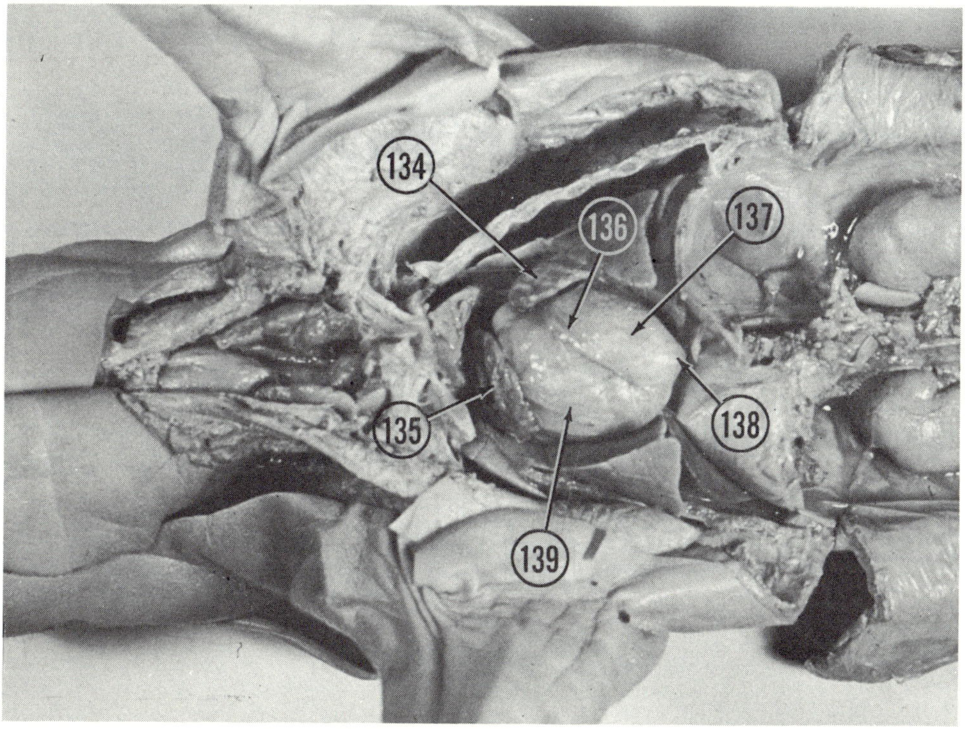
134
135
136
137
138
139

VIII. Veins

In doubly injected specimens, most of the veins contain a blue rubber compound. (Some which are not fully injected will appear brownish.) Begin just anterior to the heart. With the scalpel and scissors, slice and cut away the bone, cartilage, and muscle tissue found at the apex of a triangle formed by the ribs as they are held back from the midline. Care must be exercised to avoid cutting blood vessels. When blood vessels are found, use forceps to pick the tissue away from around them.

DISSECTION SHOWING VEINS RETURNING BLOOD FROM HEAD AND FORELIMBS

140 Internal jugular vein
141 External jugular vein
142 Innominate vein
143 Precava (anterior vena cava)
144 Subclavian vein

A Arteries

DISSECTION WITH HEART RAISED SHOWING THE MAIN VEIN RETURNING BLOOD FROM POSTERIOR REGIONS

145 Heart
146 Postcava (posterior vena cava)
147 Diaphragm
148 Larynx
149 Trachea

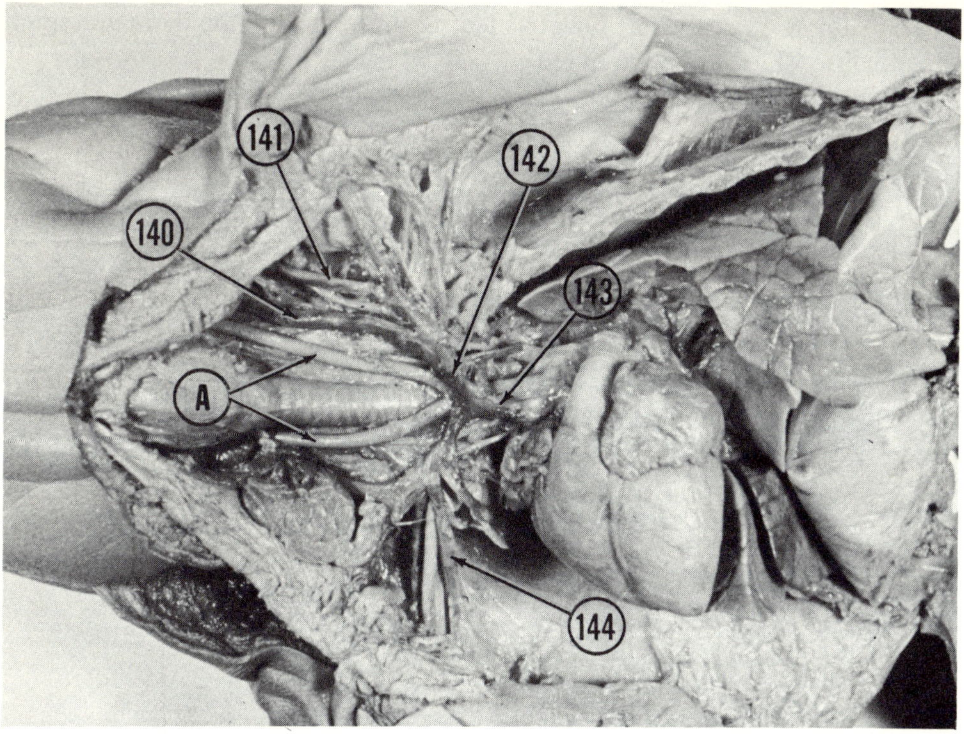
141
142
140
143
A
144

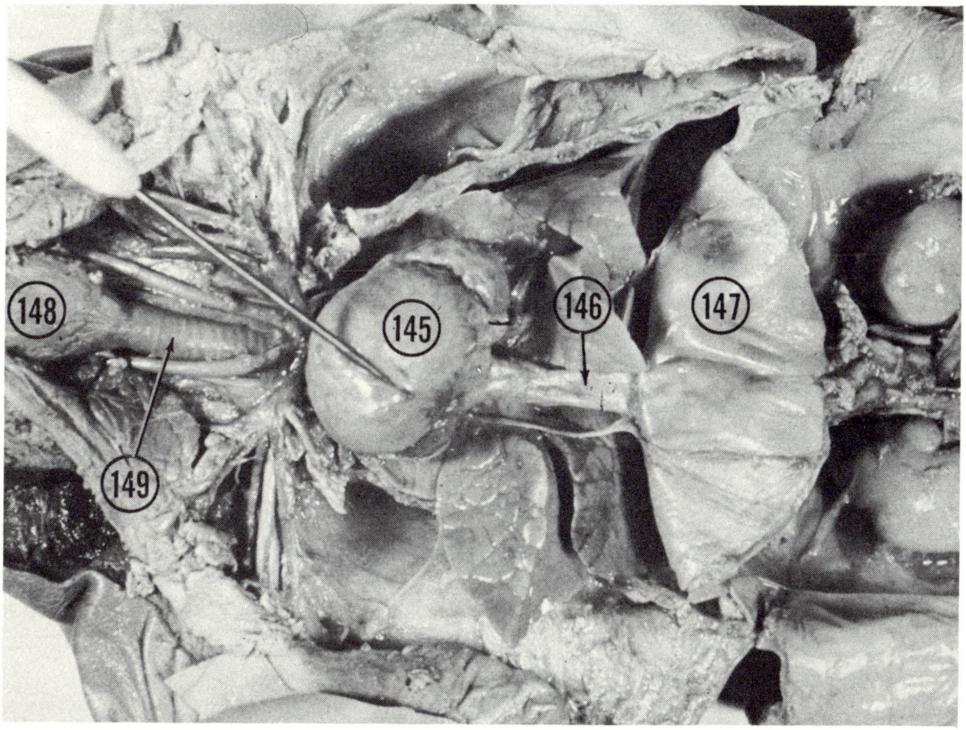
148
146
147
145
149

Remove a portion of the left lung to identify the vessels in the thoracic region listed below. The diaphragm may be removed also.

DISSECTION SHOWING VEINS IN THORACIC AND ABDOMINAL REGIONS

150 Trachea
151 Hemiazygous vein (note branches to ribs)
152 Renal vein
153 Postcava (posterior vena cava)

A Artery

The peritoneum must be removed for examination of the veins and arteries in the abdominal region.

DISSECTION SHOWING VEINS POSTERIOR TO THE DIAPHRAGM *The artery has been cut and is held up by the finger. The ureters have been cut and pulled laterally.*

154 Ureter
155 Renal vein
156 Postcava (posterior vena cava)
157 Left common iliac vein
158 Right common iliac vein

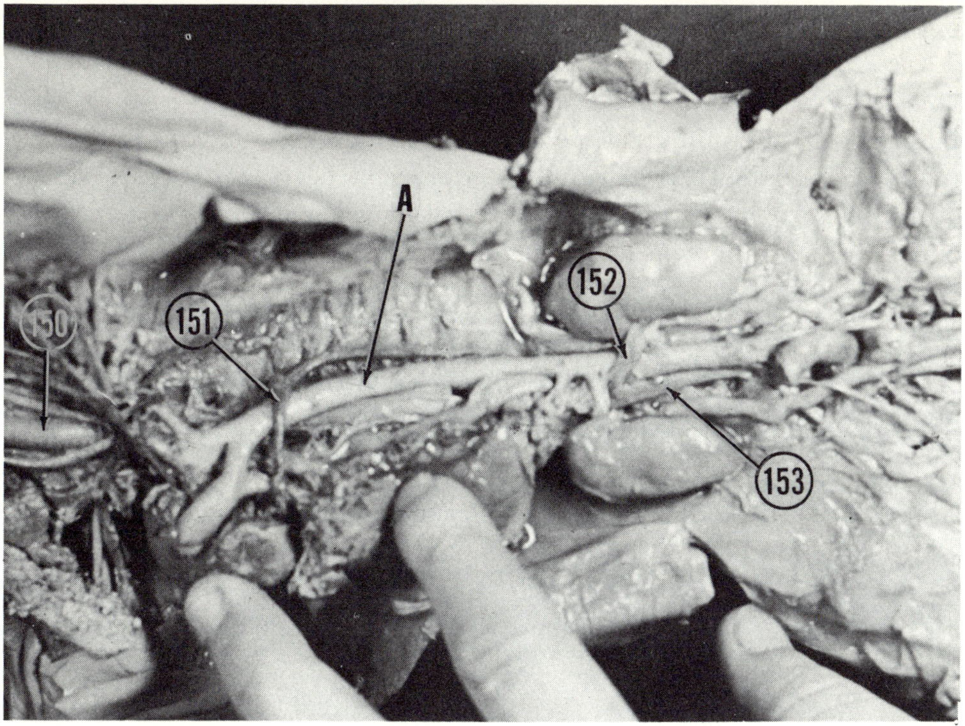
A
150
151
152
153

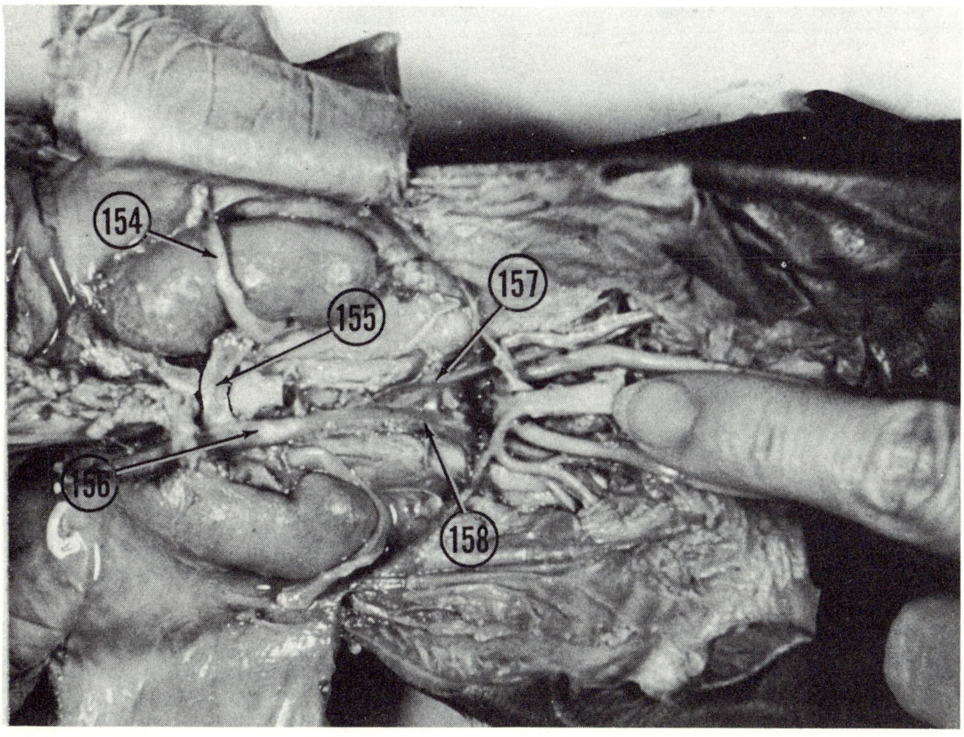
154
155
156
157
158

IX. Arteries

In doubly or singly injected specimens, the arteries are injected with a red rubber compound. Their exposure anterior to the heart will require removal of the veins in this region. Connective tissue and muscles can be carefully picked away with forceps to expose the arteries fully.

DISSECTION SHOWING ARTERIES ANTERIOR TO THE HEART *with heart pulled to right side*

159 Left subclavian artery
160 Aortic loop
161 Right subclavian artery
162 Brachiocephalic artery
163 Left auricle
164 Coronary artery (coronary vein underneath)

DISSECTION SHOWING ARTERIES ANTERIOR TO THE HEART

165 Left common carotid artery
166 Left brachial artery
167 Lateral thoracic artery
168 Ductus arteriosus
169 Pulmonary trunk
170 Right common carotid artery
171 Right thyrocervical trunk

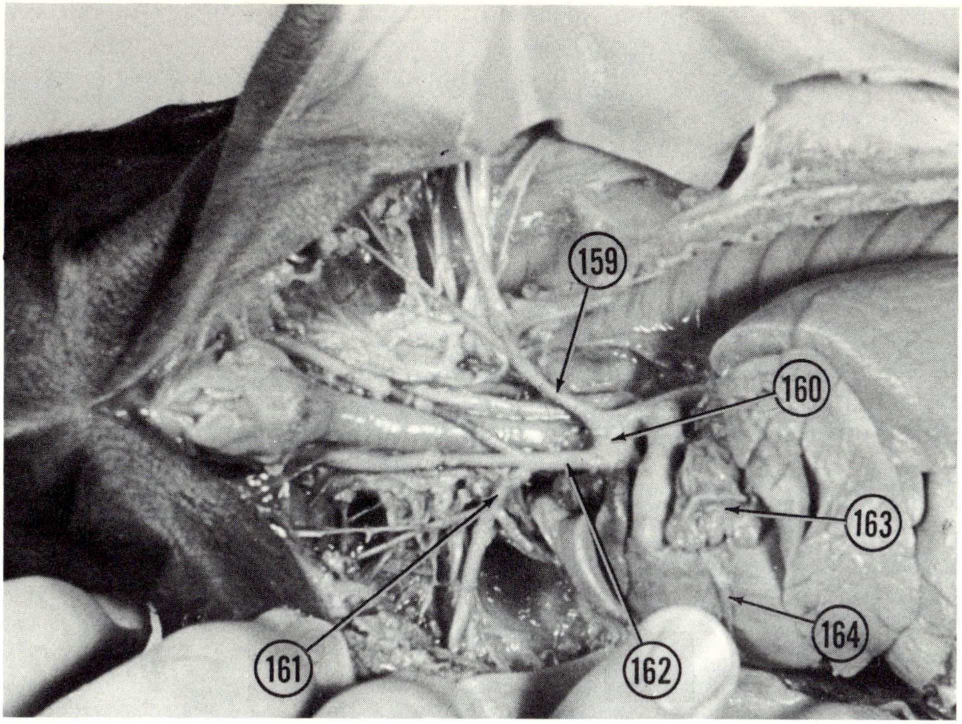
159
160
163
164
161
162

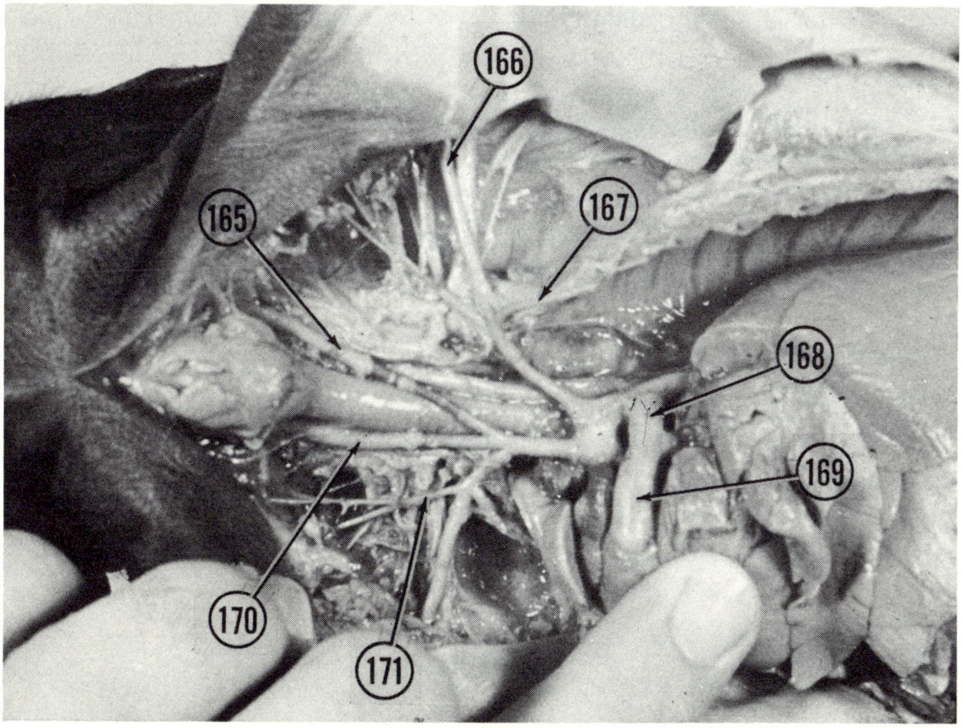
166
165
167
168
169
170
171

Remove the lungs on the left side and displace the heart to the right side for examination of the vessels seen here.

DISSECTION SHOWING ARTERIES ANTERIOR TO THE DIAPHRAGM

172 Larynx
173 Trachea
174 Esophagus
175 Aortic loop
176 Dorsal aorta
177 Pulmonary trunk
178 Left pulmonary artery
179 Ductus arteriosus

Study the vessels in the abdominal region by tracing the dorsal aorta posteriorly. It will be necessary to pick away muscle tissue to expose the external iliac arteries.

DISSECTION SHOWING ARTERIES POSTERIOR TO THE DIAPHRAGM *left ureter has been cut and pulled laterally over the kidney*

180 Dorsal aorta
181 Renal artery
182 Left external iliac artery
183 Umbilical artery
184 Coeliac artery (cut)
185 Anterior mesenteric artery (cut)
186 Right external iliac artery
187 Right genital artery

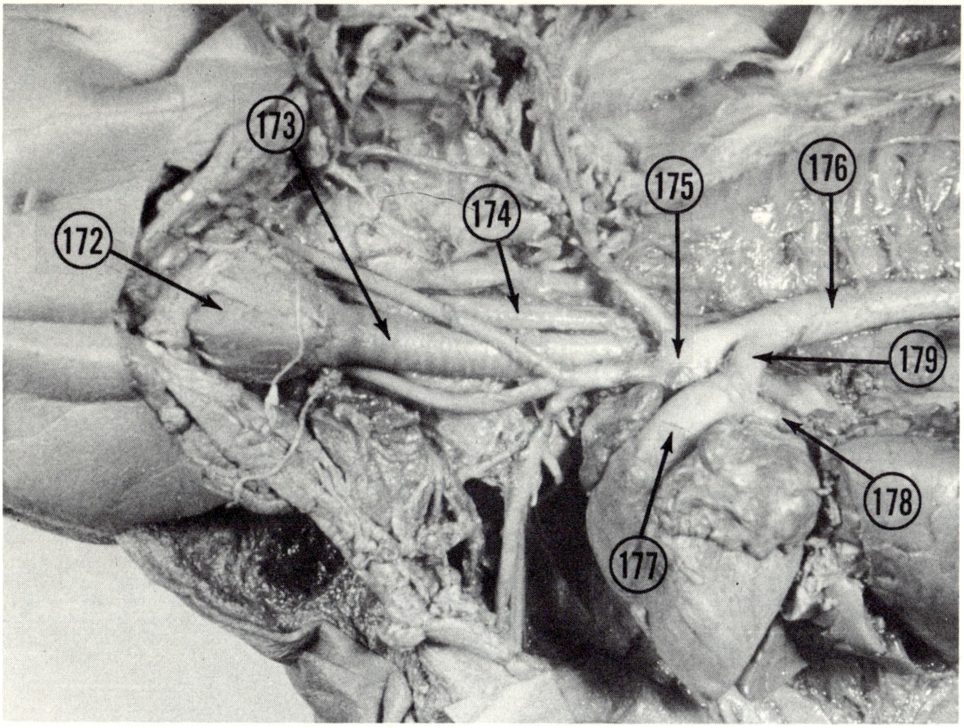
172
173
174
175
176
177
178
179

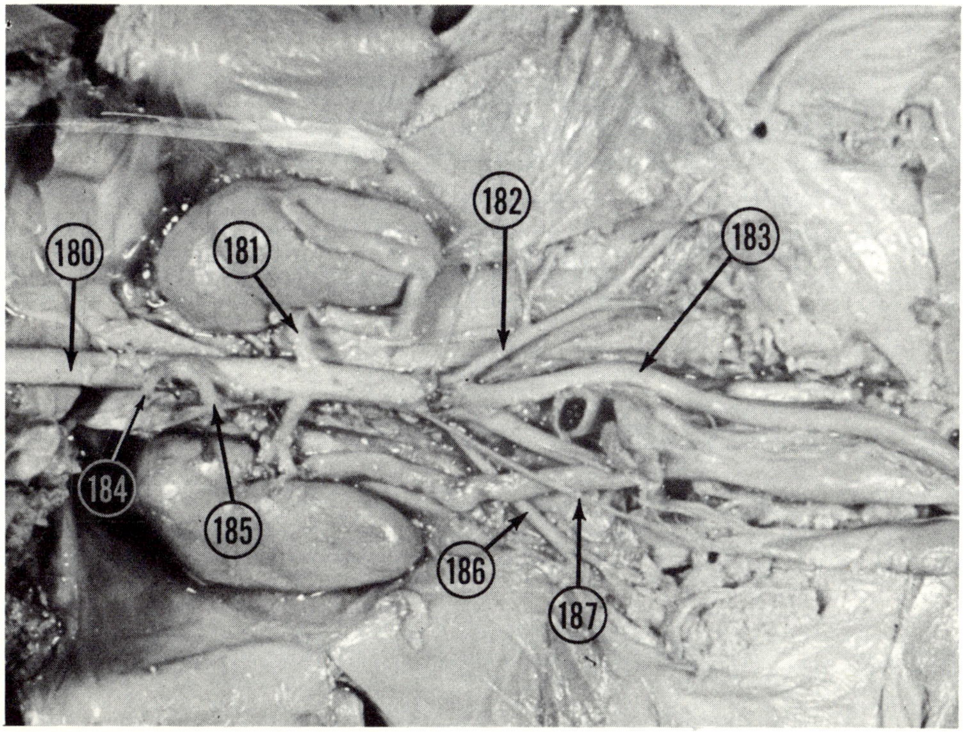
180
181
182
183
184
185
186
187

X. Brain and Spinal Cord

Sever the head from the body. With scalpel and scissors, cut through the muscles and skin as indicated in the photograph. The bone can be carefully shaved and cut away to expose the brain.

DISSECTION SHOWING DORSAL VIEW OF THE BRAIN

188 Cerebrum
189 Longitudinal fissure
190 Gyrus (ridge)
191 Sulcus (depression)

The brain is covered by two membranes. The tough outer connective tissue is called the *dura mater*. The membrane next to the brain is the *pia mater*, which is supplied with many blood vessels.

DISSECTION SHOWING DORSOPOSTERIOR VIEW OF THE BRAIN

192 Spinal cord
193 Cerebellum
194 Cerebrum

The *medulla oblongata* lies ventral to the cerebellum and is obscured in this photograph.

The complete brain can be removed for study by cutting away the bone on each side.

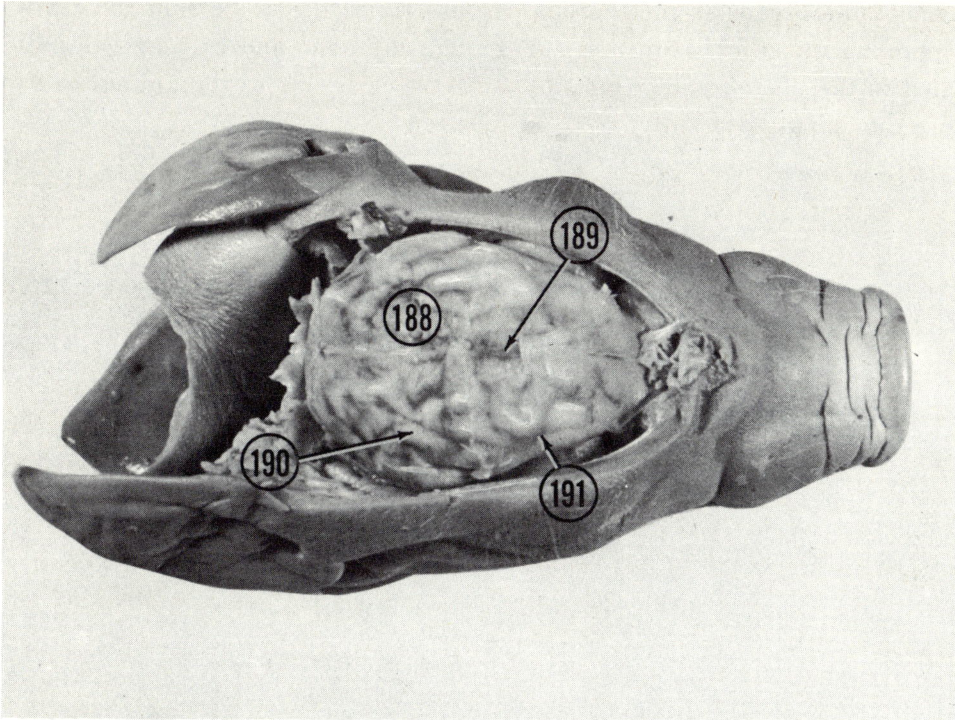
189
188
190
191

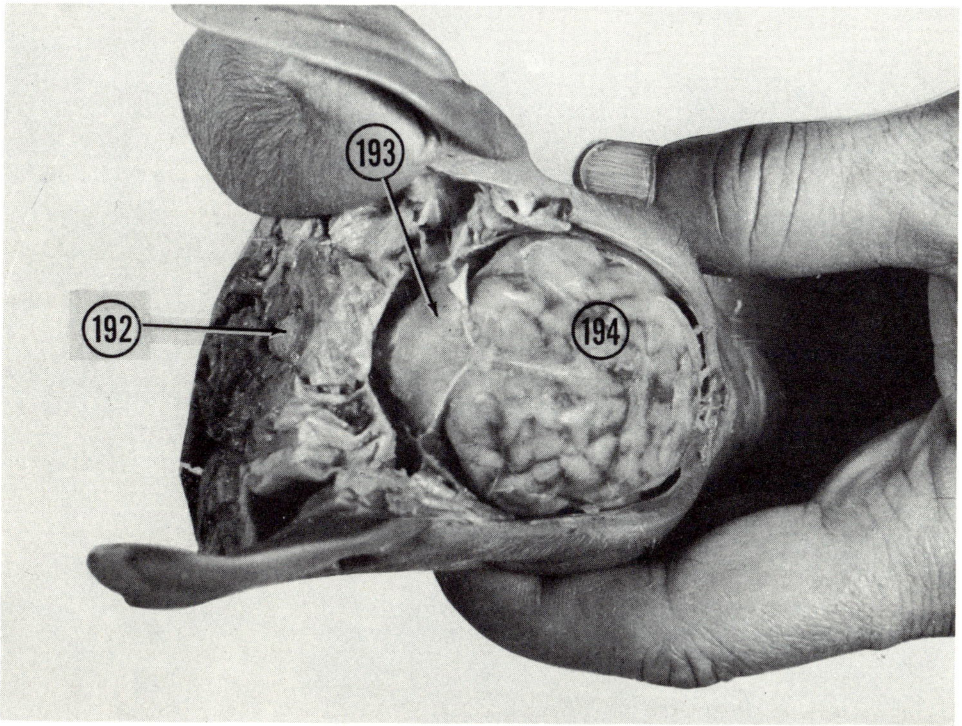
193
192
194

Remove the organs in the thoracic region. Insert the scissors in the cavity containing the spinal cord and cut through the bone and cartilage as indicated in the photograph. Apply pressure on each side of the animal to expose the spinal cord fully.

DISSECTION SHOWING THORACIC PORTION OF NERVOUS SYSTEM

195 Spinal cord
196 Nerves forming brachial plexus
197 Cut vertebra
198 Ribs

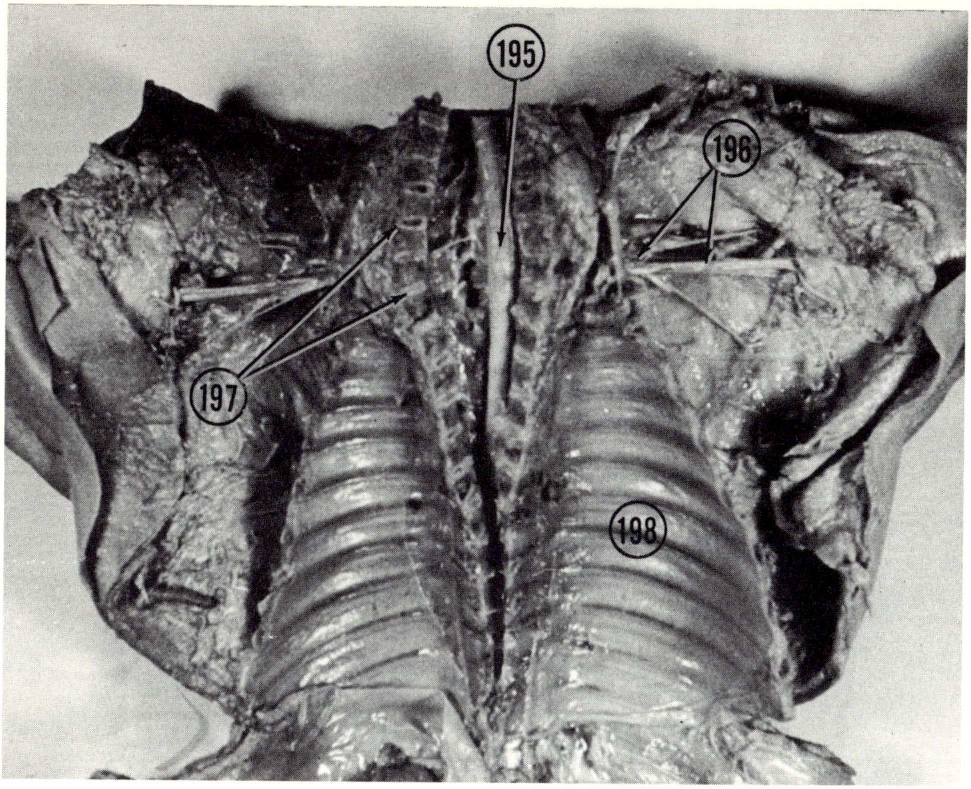
195
196
197
198

XI. Index